PRE-CALCULUS
Version 3.0

PRE-CALCULUS
Version 3.0

The Kemeny/Kurtz Math Series
True BASIC, Incorporated

Pre-Calculus

Published by True BASIC, Inc. Printed in the United States of America.

ISBN 0-939553-28-7

The Kemeny/Kurtz Math Series

The Kemeny/Kurtz Math Series aims to provide the highest quality software for high school and college mathematics topics. Each title in the series is prepared by an expert in the field, and is thoroughly tested in schools before being published.

The Series is prepared under the direction of Drs. John G. Kemeny and Thomas E. Kurtz, Professors of Mathematics and Computer Science at Dartmouth College.

True BASIC, Inc., is committed to producing the best mathematics software available, and is interested in your comments and suggestions. If you have any suggestions, please write to us at the address below.

The Kemeny / Kurtz Math Series
True BASIC, Inc.
39 South Main Street
Hanover, N.H. 03755

If you need help using Pre-Calculus, we suggest that you check this manual. It contains detailed explanations of the software. If you can't find an answer in the manual, feel free to call our Customer Support line.

603–643–3884	Customer Support Line
1–800–TRBASIC	Sales Line

Contents

Introduction	1
Setting Up the Software	5
Windows, Menus, and Dialog Boxes	9
Entering Functions	17
Entering Numbers	21
Topics	23
General	25
Angles	31
Lines	33
Triangles	35
Sine Curves	39
Conics	43
Polar Curves	47
Roots	49
Inverse Function	51
Trig Identities	53
Build Functions	55
Using PRE-CALCULUS for Self-Study	59
Saved Sessions on your Disk	65
Index	67

Introduction

PRE-CALCULUS provides a flexible tool for the teaching and learning of topics in trigonometry, advanced algebra, and elementary functions. It is not a drill or tutorial program; it allows you to select your own examples. The program is ideal for classroom demonstrations. It is equally useful for self-study, for free exploration of topics, and for the solution of problems.

For ease of use, PRE-CALCULUS is entirely menu-driven. Typing is kept to a minimum — you only need to type your formulas and a few numbers (such as intervals for plotting a function). All typed input goes through "dialog boxes," that make it easy to enter the right input and to correct mistakes.

PRE-CALCULUS contains one general-purpose tool useful throughout the course, plus ten specialized tools illustrating individual topics in a Pre-Calculus or Trigonometry course.

While the purpose of the package is to allow you to explore Pre-Calculus topics freely, a number of examples for each topic have also been provided. You may wish to try some of these examples before striking out on your own.

You can record an entire session with PRE-CALCULUS, save it, and replay it automatically.

This feature is ideal for preparing classroom demonstrations. Record them in advance, and play them during class. There are a few recorded sessions already on the disk, which you may want to try. They are listed at the end of this manual.

Overview of Topics

The topics included in PRE-CALCULUS are:

General	General-purpose tool for plotting and evaluation
Angles	Illustrates angles and trigonometric functions
Lines	Equation of a straight line
Triangles	Solves triangles given SAS, ASA, SSA, SSS
Sine Curves	Illustrates the general function A*sin(B*x+C) with varying parameters
Conics	Illustrates conic sections
Polar Curves	Graphs with polar coordinates
Roots	Finds roots of a given equation
Inverse function	Illustrates the concept of the inverse of a function

Trig identities	Verifies trigonometric identities
Build functions	Composition of functions

Setting Up the Software

This chapter describes how to get started using PRE-CALCULUS on your computer. Read this chapter before you start to use PRE-CALCULUS.

Requirements

PRE-CALCULUS runs on a variety of microcomputers. The minimum requirements for each brand of computer are listed below.

Computer	**Requirements**
IBM PC or PS/2	256K and a graphics adapter
Apple Macintosh	512K
Commodore Amiga	512K
Atari ST	512K

You don't need any software other than your operating system software and the PRE-CALCULUS disk to run the Pre-Calculus package.

Make a Back-up Copy!

Because disks are so easily damaged, we suggest that you make a working copy of your disk. To do so, follow the disk copying instructions which came with your computer.

After you've made a copy of the disk, it's a good idea to put the original away in a safe, dry place, and to use it only if your copy is damaged or destroyed.

Note: While we encourage you to make a copy of the disk for your personal use, we caution you that you are not permitted to make copies of the software to distribute to others, and that distributing copies constitutes software theft.

Using Hard Disks or Networks

PRE-CALCULUS works fine on a hard disk or networked file system. Just follow your computer's instructions for copying files to a hard disk.

Be sure to create a new directory (or "folder") for PRE-CALCULUS. Put all the files from the PRE-CALCULUS disk in this folder.

Reminder: If you want to share one copy of PRE-CALCULUS across a network, please call True BASIC, Inc., to arrange for a site license.

Printing a Screen

Every computer has a different way to print its screen on a printer. Usually you must press some combination of keys on the keyboard, and the computer then "takes a snapshot" of the screen onto the printer.

Many computers require some kind of special preparation before you can print a picture on your printer. Often you must run some utility program before you start running PRE-CALCULUS. Read your computer's *Owner's Manual* to find out what you must do.

Starting PRE-CALCULUS

Here's how to start PRE-CALCULUS on your computer.

1. Turn on the computer and use your Operating System disk to get the computer running.
2. Put the PRE-CALCULUS disk in the disk drive.
3. If you're using a DOS operating system on an IBM PC or compatible, type PRECALC at the A> prompt.

   If you're using an icon-oriented system, double click on the icon labelled PRE-CALCULUS.

In a moment you'll see the PRE-CALCULUS windows.

Windows, Menus, and Dialog Boxes

This chapter describes the windows, menus and dialog boxes used in PRE-CALCULUS.

Windows

PRE-CALCULUS uses three windows: one at the very bottom of the screen to display messages, one at the left of the screen for text displays, and one at the right of your screen for plotting.

In many instances, you can plot more than one curve on the same set of axes. When you do this, the program draws each new function in a different line style or color. The differences in line styles depend on the computer you're using.

Note: *Since PRE-CALCULUS runs on many kinds of computers, the figures in this manual may not exactly match what you see on your screen. However, the overall organization and use is the same on all computers.*

Menus

You control the package by menu selections. On some computers you use the mouse to make menu selections. On others you use function keys.

When you begin, you'll first see an introductory screen, then the main menu bar, which offers the following choices:

Topics Session Units Help

The Topics menu offers twelve choices. The first, General, selects the general purpose tool. The next ten are for the special purpose tools. Use the final selection, Quit, when you're ready to end your session.

When you select a topic, the menu bar changes to show menus appropriate for that topic. When you choose the option Exit under a topic, you return to the main menu.

Help and Glossary menus

Each topic's menu bar contains a Help menu and a Glossary menu. These two menus give more information about that topic.

The Help menu gives explanations of that topic's menus.

The Glossary menu shows a list of terms for which the software will provide definitions. For example, if you select Roots from the list of glossary terms, the program will print the definition of roots in the message window.

Units menu

The Units menu lets you switch between radians and degrees for your work. PRE-CALCULUS is preset to use radians, but you can switch to degrees at any time by choosing the Units menu, then selecting Degrees.

Session menu

PRE-CALCULUS lets you save your work during the run of the program for later playback for demonstration or review. These saved sequences of examples, or "sessions," can be especially helpful in organizing classroom demonstrations.

The Session menu offers four options:

Play a session

Record a session

Stop recording

Quit

Choose Play a session to replay a sequence which is saved on your disk. We have provided several sessions for you; they are listed at the back of this manual. You may find that playing a saved session is a good way to become acquainted with the features of the software. We suggest trying the sessions GENERAL and COMPOSE.

Choose Record a session when you want to save your own work. After choosing that option, just run the program normally. When you have finished with the work you want to save, return to the main menu bar and choose Stop recording from the Session menu. You'll be asked to provide a name for your recorded session. Your session will then be saved on your disk under the name you gave, and you can later replay that session.

Select Quit from the Session menu to stop running PRE-CALCULUS.

Menu Selections on an IBM PC

To choose from the main menu bar on the IBM PC, press the function key that corresponds to your choice: F1 for a list of topics, F2 for Session options, and F10 for Help.

After you choose one of the menu bar items, you'll see a menu with the choices for that item. Use the up/down cursor keys, then press Enter when your choice is highlighted. You can also use the Home key to move to the first item in a menu, and the End key to move to the last item.

If you select a menu by mistake and don't want any of the items listed in that menu, press Esc to escape.

Menu Selections on Other Computers

On a Macintosh, Amiga, or Atari ST, just use the menus as standard pull-down menus. After you choose one of the main menu items, you'll see a new menu bar with the choices for that item.

Dialog boxes

You provide input to PRE-CALCULUS by entering functions and numbers in a dialog box, or input box.

To edit the functions in this box, simply move the cursor to the place where you want to make a change or correct an error, delete any incorrect characters, and type the correct characters.

Each dialog box also has a legend or buttons at the right with three or four options. The standard ones are:

OK	Dialog box is correct; proceed.
Help	Explain how to use this dialog box.
Cancel	Abandon this input.

The first dialog box presented under a new topic may also offer Example as a choice, in case you want to try a built-in example rather than entering your own. You may also select an example by choosing Pick an Example from the Options menu.

Editing a dialog box

On the Macintosh, Amiga, or Atari ST, the dialog boxes are standard dialog boxes. Use them in the usual way. If you're not sure how to use dialog boxes on your computer, consult your computer's *Owner's Guide*.

On an IBM PC or compatible, use the following keys for editing:

Right arrow	Move cursor one character to the right
Left arrow	Move cursor one character to the left
Down arrow or Tab	Move cursor to next field
Up arrow or Shift tab	Move cursor to previous field
Home	Move to the first field in the dialog box
End	Move to the last field in the dialog box
Backspace	Delete character to the left of the cursor
Ctrl-Backspace	Delete a complete line in the dialog box
Del	Delete character at the cursor position

Your previous input may show in a dialog box. You can edit this input, or you can delete the complete entry to begin from scratch.

Entering Functions

The package is designed for functions of one variable, $y = f(x)$.

Several of the units ask you to enter a function f. You define this function f in terms of a variable x, constants, and arithmetic operators. Use only the letter x or the letter t to represent the variable in your expression — if you use another letter or combination of letters, the program will tell you that you used an illegal symbol and the dialog box will reappear so that you can edit the function.

You may use the following arithmetic operators in defining your function:

+	Addition
-	Subtraction
*	Multiplication
/	Division
^	Exponentiation, or raising to a power. (Thus 10^2 = 10^2 = 100.)

Use parentheses as needed for determining the order of operation or for clarity. You may also include blanks in the expression.

You may include the constant *pi* in your function definition.

Available Functions

The following functions are also available.

sin(x)	sine of x
cos(x)	cosine of x
tan(x)	tangent of x
cot(x)	cotangent of x
sec(x)	secant of x
csc(x)	cosecant of x
asin(x)	arcsine of x
atn(x)	arctangent of x
abs(x)	absolute value of x
sqr(x)	square root of x
log(x)	natural logarithm of x
exp(x)	e^x
int(x)	greatest integer function

You may enter the function names and the constant *pi* in any combination of upper and lower case letters.

Examples

The following examples are legal responses to the prompt for a function:

sin(*x*)^2

x^7 + 3***x*^5 - 5***x*^3 + 6***x*

(*x*^2 + 1)/(3***x* + 2)

1/exp(*x*)

The following expressions are not legal for the reasons given.

2*t* + 5	Needs operator between 2 and *t*: should be 2***t* + 5
6***y*^2 - 7	Can't use *y* as the variable: should be *x* or *t*
abs((*x*+3)/(*x*^2-5)	Parentheses don't balance

If you enter an illegal function, the program will display your input in the Message window, pointing to the problem, with an appropriate error message. You can then edit your function in the dialog box and try again.

Entering Numbers

You may enter constants as integers, fractions, decimals, or in exponential notation. Some examples of numeric constants are:

546

2/3

3*pi*/4

-23.4567

1.067e4 (1.067 times 10^4)

7.654e-4 (7.654 times 10^{-4})

You may enter multiples of *pi* as 2*pi*, 3*pi*/2, etc. With multiples of *pi*, you may omit the multiplication sign (*).

You may not include commas, spaces, or other special characters in a numeric constant. The following examples are not legal numbers:

345,678

#45

5 678

99.6%

Topics

The following sections describe each of the individual tools available in PRE-CALCULUS.

You can read these sections in any order. Since the General menu is the most commonly used, we suggest that you read its section first.

TOPICS

General

Use this topic to explore the behavior of functions through graphs and tables of values.

General starts by asking for a function. Then you have a number of choices. You may print tables of values for the function. You may plot several functions in the same picture (with varying colors or line styles used to distinguish them).

After you choose a function, either by entering your own, by modifying one already in the dialog box, or by picking an example, you'll see the following menu bar at the top of the screen:

Options Plot Units Glossary Help

The Plot menu offers the following options:

Plot all	Draw the graph each time you enter or select a new function.
Don't plot	Don't graph the functions.
Rescale	After a plot, change the coordinates to make the graph more attractive.

Plot all and Don't plot are toggle options — one or the other (checked on the menu display) is in effect.

Options Menu

The Options menu provides the following choices:

New function	Supply formula for a new function.
Pick an example	Choose an example from a list.
Evaluate (Table)	Print a table of values of the current function. You will be asked to provide values for From, To, and Step via a dialog box.
Print formula	Display the formula of the current function.
Clear	Clear the screen. Use it when you want to plot on a new set of axes instead of the same axes as the previous function.
Exit	Return to the main menu.

With the Evaluate option, a response of -2, 3, and .1 will print values from $x = -2$ to $x = 3$ in intervals of .1. If you provide values for From and To only, a step of 1 is assumed. If you enter only one number, a single value is printed.

You may express the values as constants or multiples of *pi*. If you choose the Evaluate option by mistake, you can return to the menu of options by choosing the Escape or Cancel option.

If you request a table that contains more values than will fit on a single screen, the program will display one screenful of the table, then give you the choice of continuing or quitting.

If you want to see more values of the table, choose the Continue option; otherwise choose the Cancel option.

Examples

Try some examples. From the main menu, choose the topic General. When you're asked to enter a function, choose Example. Choose the example in the list called "sin, cos" and watch the playback. When the example is finished, you'll see the message "Example completed" in the Message window.

After you have observed the example, try some experimentation on your own. First choose the option Clear, then New function.

The dialog box will display the function from the last session. Delete that function and enter:

x^3 - 3**x*^2 + 2

While you're still in the dialog box, you can choose the Help option to see what input is required.

Be sure to press the Return key after you type the complete formula and are ready to move on. If you make any errors in entering the formula, the program will display a message in the message window and indicate where the error occurred. You can then correct the error in the dialog box.

The toggle option Plot all will still be in effect from the example, so you'll be asked to supply an interval for plotting. Enter -3 and 3 in the dialog box.

Next display a table of values for the function. Choose Evaluate from the Options menu. In the dialog box requesting values of From, To, and Step, enter -8, 8, and 2. You'll see the values of x and $f(x)$ from x = -8 to x = 8 in increments of 2.

Now plot another function on the same set of axes. What if the first term were x^4 instead of x^3? Choose New example from the Options menu, then edit the function in the dialog box so that the function is

x^4 - 3*x^2 + 2

After the plot of the second function, you may want to choose Rescale from the Plot menu to redraw the graphs with more appropriate y-coordinates. See Figure 1.

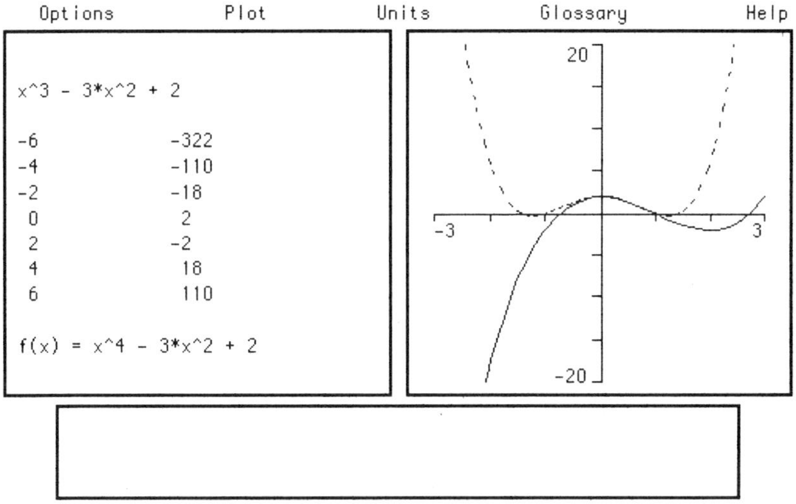

Figure 1

Choose Exit from the Options menu when you're ready to return to the main menu.

TOPICS

Angles

Use this topic to examine angles and the six trigonometric functions.

Angles asks for the *X*- and *Y*-coordinates of a point, then draws the circle with origin 0,0 and the point on its circumference.

It computes and displays the distance of that point from the origin and draws the angle created by a line from the origin to that point and the *X*-axis, then computes that angle and its six trigonometric functions.

As an example, enter 12 and 5 when asked for the coordinates of a point. If you're working in radians, the value of the angle will be expressed in radians, or if you're working in degrees, it will be expressed in degrees.

Try another example by choosing New angle from the Options menu. This time, enter 12 and -5 as your coordinates. You'll see the result shown in Figure 2.

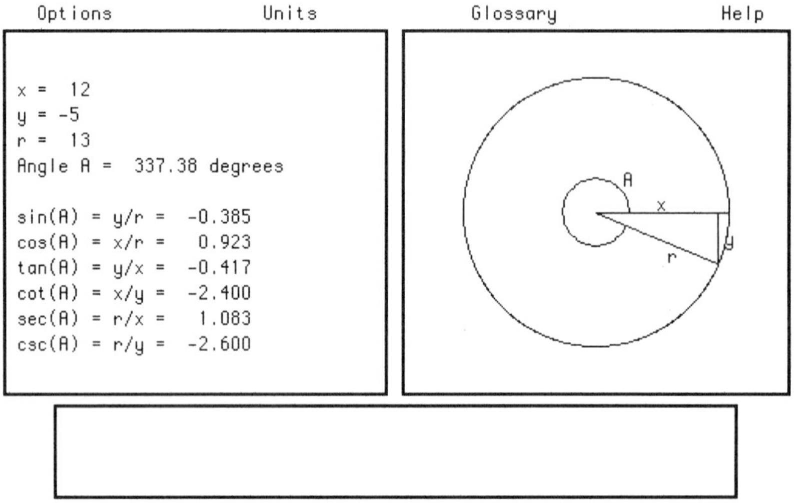
Options
Units
Glossary
Help
x = 12
y = -5
r = 13
Angle A = 337.38 degrees
sin(A) = y/r = -0.385
cos(A) = x/r = 0.923
tan(A) = y/x = -0.417
cot(A) = x/y = -2.400
sec(A) = r/x = 1.083
csc(A) = r/y = -2.600
A
x
y
r

Figure 2

TOPICS

Lines

Use this topic to examine the concept of the slope of a line and relationships between straight lines and coordinates in the X-Y plane.

Lines determines the equation of a line in the form $y = m*x+b$. You can supply either:

two points on the line;

one point and the line's slope.

Use the Options menu to choose which of these input forms you want.

The Lines Options menu looks like this:

Two points

Point and slope

Pick an example

Exit

If you choose Two points, you'll be asked to enter the coordinates of two points. The program will then print the equation of the line, plot the line, and print the values of the X- and Y-intercepts and the slope.

If you choose Point and slope, you'll be asked to enter the coordinates of one point and the slope of the line. Again, the program will print the equation of the line, plot the line, and print the *X*- and *Y*-intercepts.

Try an example giving two points on a line. Enter 2 and 4 as the coordinates of the first point, and 4 and -2 as the coordinates of the second. You'll see the result shown in Figure 3.

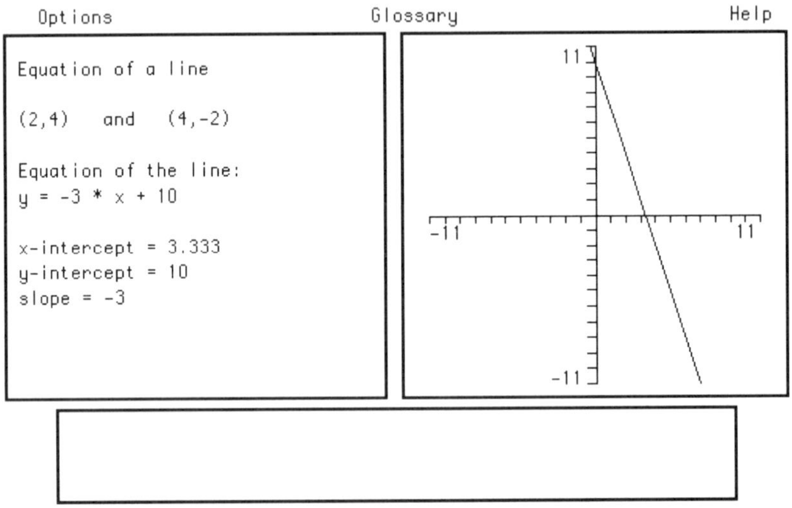

Figure 3

TOPICS

Triangles

Use this topic to solve triangles based on given values of sides and angles. Observe some combinations of sides and angles that define more than one triangle, and some that cannot define a triangle.

Triangles solves triangles given:

SAS	two sides and the included angle,
ASA	two angles and the included side,
SSS	three sides, or
SSA	two sides and an adjacent angle.

The Options menu lets you choose which type of input you wish to give Triangles. After you choose from the menu, you'll be asked to enter values for the appropriate sides and angles. Triangles then draws the triangle and prints the values for all sides and angles. It always reminds you whether you are working in degrees or radians.

As an example, choose SSS. When you're asked for values for the sides, enter the values for the famous 3-4-5 right triangle:

Side = 3
Side = 4
Side = 5

You'll see the results shown in Figure 4.

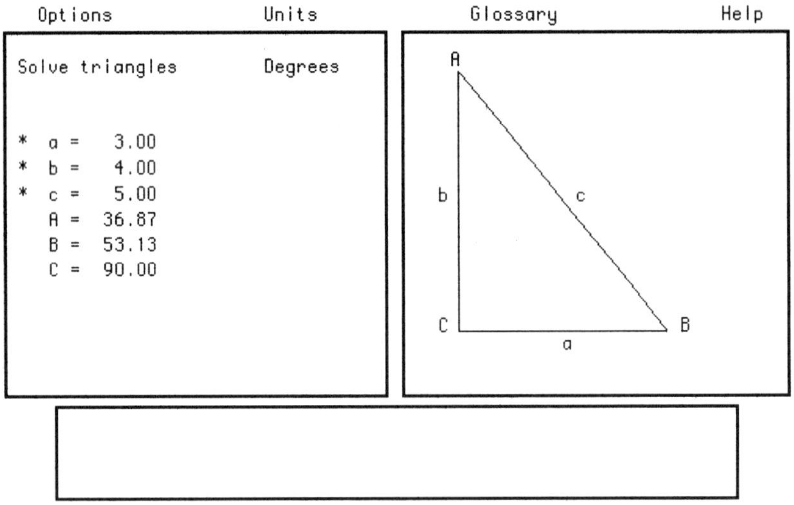

Figure 4

Now choose SSS again and enter the values 3, 4, and 8 when asked for the values of the sides. The program will tell you that it's not possible to define a triangle with sides of those lengths.

Now choose SSA. Enter the following values:

Side = 40
Side = 30
Angle = 30 (*pi*/6 if you're using radians)

You'll see that the program handles two cases, and illustrates two different triangles defined by that particular SSA input. After the display of the first triangle, you'll see a message that says there is another case.

Press any key or click the mouse and Triangles will display the other triangle.

Choose Exit to return to the main menu bar.

TOPICS

Sine Curves

*Use this topic to explore the general function A*sin(B*x+C) and how the function changes when you vary the parameters A, B, and C.*

Sine curves lets you choose different parameters to examine the general function A*sin(B*x+C). It graphs multiple functions on the same set of axes so you see what happens as you change A, B, or C.

You'll first be asked to define a function by entering the three parameters A, B, and C. Then it graphs each function that you enter. The menu gives you the following choices:

New function

Pick an example

Rescale

Clear

Exit

The New function option lets you define a new function by changing any of the parameters A, B, or C. In many instances, you'll just want to vary one of the parameters at a time, to examine the effect a change in that parameter has on the function.

After you have defined and graphed several functions on the same set of axes, you'll probably find that the original plotting coordinates are not appropriate for all of the functions. Use the Rescale option to change coordinates for graphing; it works just as the Rescale option in the General menu does.

Choose Clear if your graph becomes too cluttered. If you want to begin a new example with a new family of functions, choose Clear to erase the graph.

The Exit option returns you to the main menu bar.

Examples

As an example, first enter 1, 1, and 0 as A, B, and C respectively. That will define the function 1*sin(1*x+0), or simply sin(x). The next dialog box will ask for the plotting interval; enter *-pi* and *pi* if you are using radians or -180 and 180 if you are using degrees.

Now see what happens when you change C. Choose the option New function and edit the value of C so that the new value is *pi*/4 (45 degrees if you are working in degrees). After you see that function, choose New function again, and enter *-pi*/2 (or -90) as the value of C. You'll see that changing C moves the sine wave to the left or right, but doesn't affect its height or period.

Now examine the effect on the function of changing the parameter B. Choose Clear, then New function. Leave the parameters unchanged, and when the dialog box asks for the plotting interval, enter *-pi* and *pi* (or -180 and 180). After the plot, choose New function again, and change B to 2. You'll see the effect of changing the value of B on the period of the function.

Now see what happens when you change A. Choose Clear and New function again. Enter 2, 1, and 0 to define the function 2*sin(x). After the plot, choose New function again and change the value of A to -1. See Figure 5.

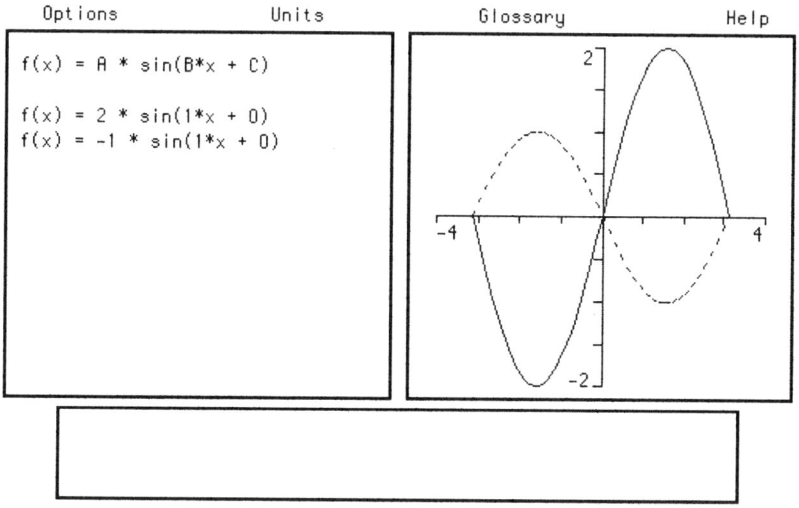

Figure 5

TOPICS

Conics

Use this topic to graph conic sections and examine what effect the coefficients of the equation have on the shape of the conic.

Conics illustrates conic sections. You'll be asked to enter the coefficients A, B, C, D, E, and F for a relation of the form

$$A*x^2 + B*x*y + C*y^2 + D*x + E*y + F = 0$$

Conics will determine whether the relation is an ellipse, a parabola, or a hyperbola, and draw its graph. Conics determines the appropriate coordinates.

Try the saved examples; they illustrate a variety of cases. Then move on to do some experimentation on your own.

Figure 6 illustrates the ellipse defined by A = 1, B = 3, C = 4, D = -2, E = 18, and F = 1.

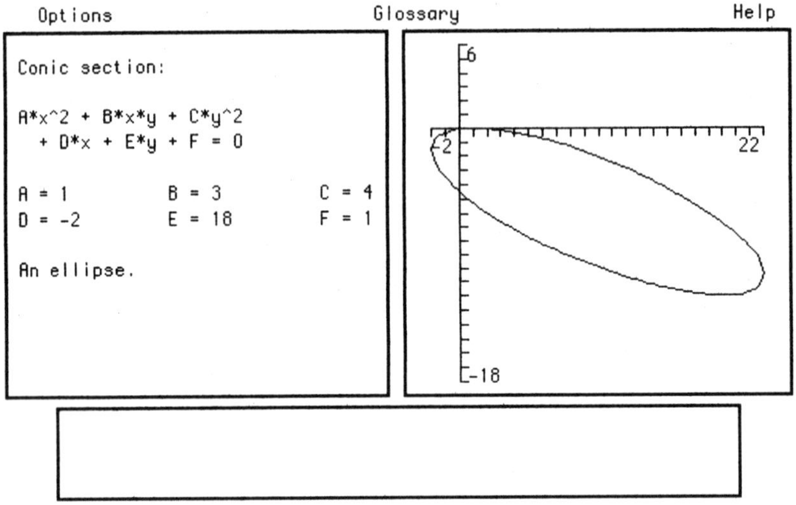

Figure 6

Figure 7 illustrates the hyperbola defined by A = 1, B = -2, C = -4, D = 2, E = 16, and F = -19.

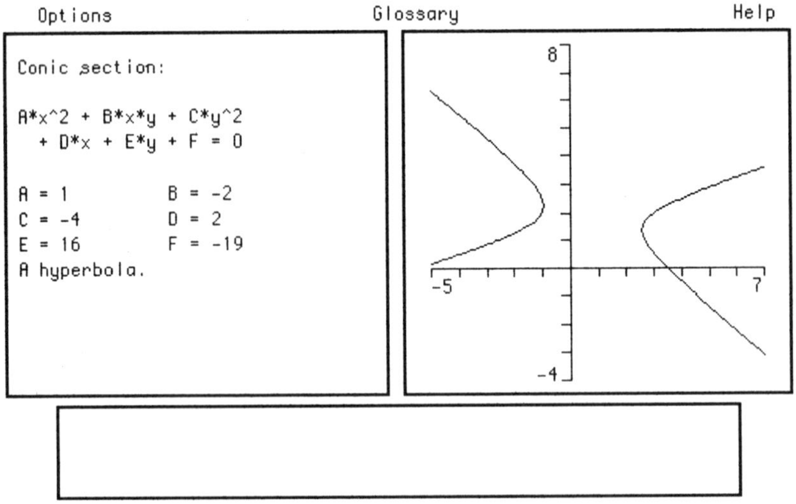

Figure 7

Conics also handles "degenerate" cases that produce graphs of exactly one point, a single line, parallel or crossing lines. Enter the coefficients 1, 0, -4, -6, -4, and 8 in response to the dialog box, and you'll see that the resulting graph is a pair of crossing lines.

TOPICS

Polar Curves

Use this topic to graph polar curves.

Polar Curves draws curves in polar coordinates. The graph is the set of points *(f(t),t)*, where *t* is an angle and *f(t)* is a specified function. Polar Curves plots the function $r = f(t)$, where

$$x = r*\cos(t) \quad \text{and}$$
$$y = r*\sin(t)$$

You'll be asked to enter the function *f(t)* and the beginning and ending values of the angle *t*. Then you'll see a graph of the curve *(f(t),t)* in polar coordinates.

Enter cos(2**t*) when prompted for *f(t)*, and 0 and *pi* when asked for the interval. Then choose the option New interval and enter 0 and 2*pi* when asked for the values of "From" and "To." You'll see the result shown in Figure 8.

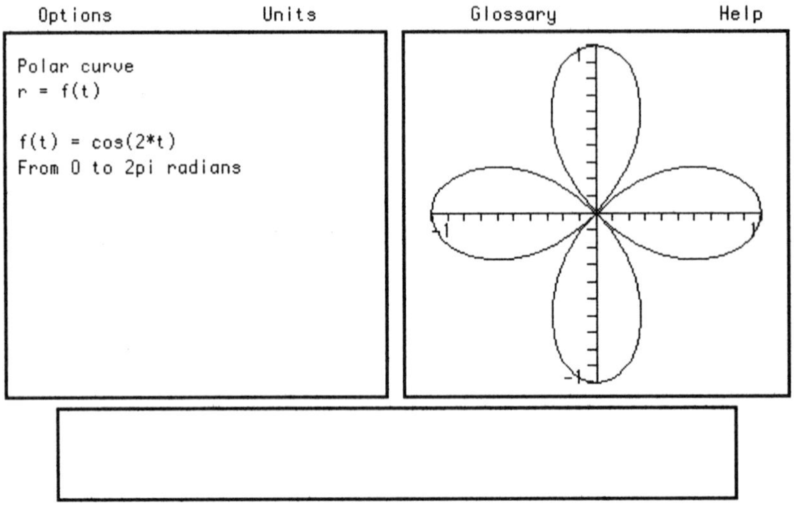

Figure 8

The saved examples illustrate a variety of interesting polar curves.

TOPICS

Roots

Use this topic to find the roots of an equation, observe the roots graphically, and see the graphic representation of the algorithm used to find the roots.

Roots finds the real roots of a given function within the interval you choose. You'll first be asked to enter a function in the dialog box.

Enter a function just as in the General menu. You'll next see a dialog box requesting the endpoints of an interval. Enter the smallest and largest values of *x* in your chosen interval. The program will then find and print the roots of the function in that interval and will plot the function.

As an example, enter

x^4 - 4**x*^2 + 2

in response to the request for a function. Enter -3 and 3 as the endpoints of your interval.

Next try a trigonometric function. Use radians as your units. Enter sin(*x*) - cos(2**x*) when asked for the function *f(x)*. Enter *-pi* and 3*pi*/2 when asked for the endpoints of the interval.

After Roots displays the roots, choose "In terms of *pi*" from the Options menu. That will display the values of the roots in multiples of *pi*. See Figure 9.

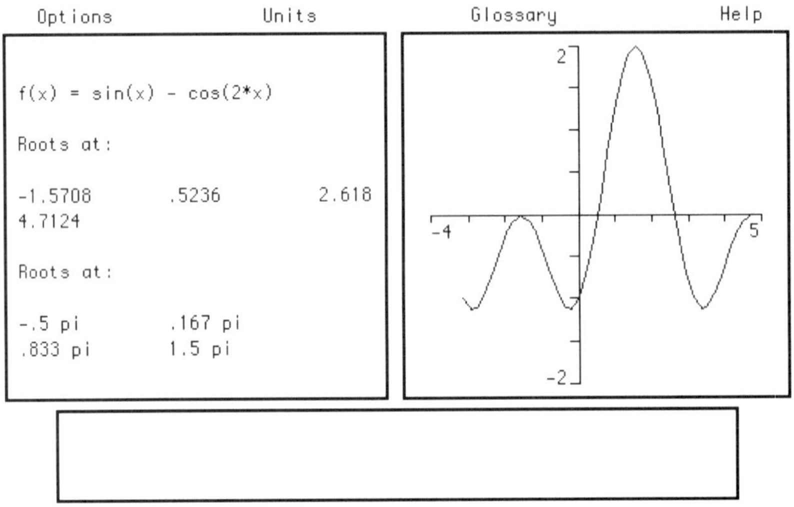

Figure 9

Roots also offers the option "Find graphically." When you choose this option, Roots will illustrate how it searches for roots by highlighting the segments where it searches for a change in sign. You'll be asked to touch any key or click the mouse to show each smaller interval as Roots closes in on a root. Try the saved example "graphic."

TOPICS

Inverse Function

Use this topic to graph a function and its inverse on the same set of axes, explore the concept of inverse, and determine when a function does or does not have an inverse in a given interval.

Inverse Function plots a function and its inverse. Enter the function as usual in response to the request in the dialog box.

You'll be asked to enter coordinates in response to "From" and "To" in the dialog box. Enter the two x-coordinates which determine the interval of the function in which you're interested.

The function itself will be plotted in one color or line style, and its inverse in a different color or line style.

As an example, enter x^3 when asked for the function, and plot it from -2 to 2. You'll see the result shown in Figure 10.

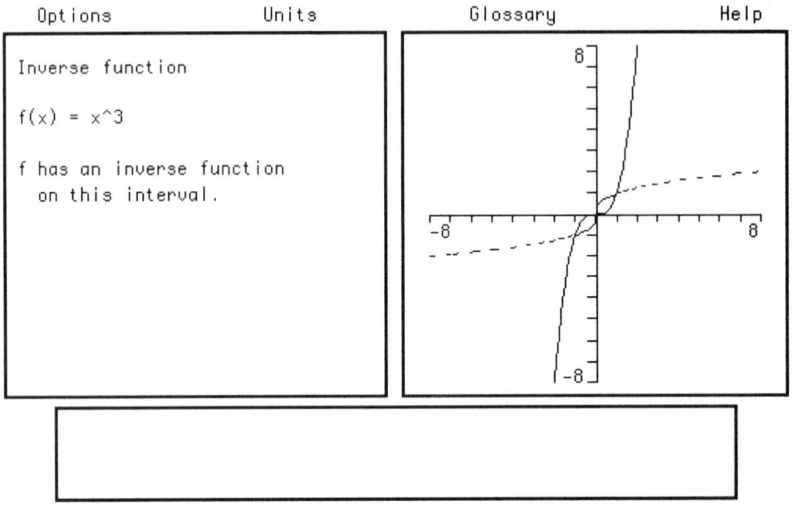

Figure 10

If the function has no inverse in the specified interval, you'll see the following message:

F does not have an inverse on this interval.

Try a different interval.

Choose New interval from the Options menu to try a different interval.

TOPICS

Trig Identities

Use this topic to verify trigonometric identities and to determine some identities on your own.

Trig Identities lets you verify trigonometric identities. The dialog box will ask you to enter two functions. Trig Identities then examines these functions from -2*pi* to 2*pi* radians or -360 to 360 degrees.

If the functions are equal in that interval, Trig Identities tells you that the equation made by setting the two functions equal to each other is indeed an identity. If not, it tells you that the equation is not an identity. If the identity is valid only in certain quadrants within the interval, it prints the intervals in which the equality holds.

If the identity holds, Trig Identities plots the function. If it is not an identity or if the identity is valid only in certain quadrants, it plots both functions.

Pick a few examples from the list of saved examples. Try the double angle formula, given in the list as sin(2*x).

Next try an example which is a valid identity only in certain quadrants. Pick the second example in the list (cos(x/2) +). See Figure 11.

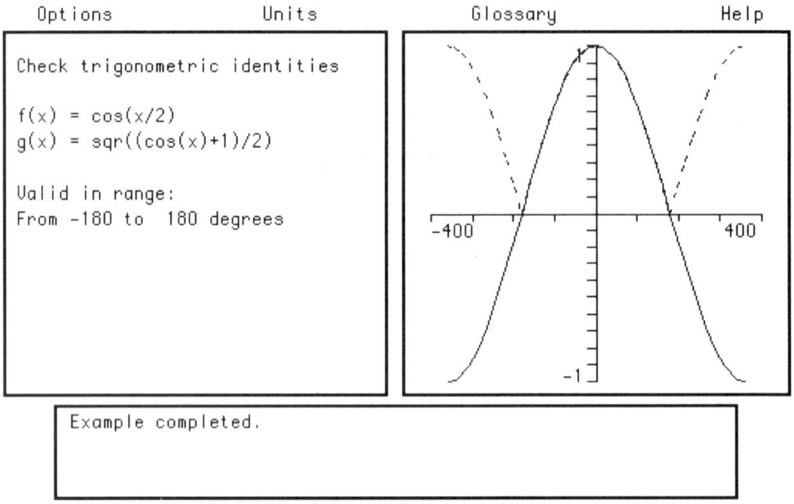
Options
Units
Glossary
Help
Check trigonometric identities
f(x) = cos(x/2)
g(x) = sqr((cos(x)+1)/2)
Valid in range:
From -180 to 180 degrees
-400
400
1
-1
Example completed.

Figure 11

TOPICS

Build Functions

Use this topic to create complicated functions from simpler ones, and to examine composition of functions.

Build Functions lets you combine and compose new functions from existing ones. You can have up to ten active functions at any one time.

When you create a new function, it will be given a name such as *f1*, *f2*, and so on, up to *f10*. You can use those names to refer to existing functions when you build a new function in the dialog box.

The Options menu offers the following choices:

New function Enter a new function. You may define the function from scratch, just as in the General menu, or you may build new functions from existing ones by using the names *f1*, *f2*,... to refer to existing functions. For example, if $f1(x) = \sin(x)$ and $f2(x) = \cos(x)$, you can define a new function, such as $f1(x)+f2(x)$ or $f1(f2(x))$.

Pick an example	The usual example option which lets you pick a saved example from a list.
Plot	Plot the current function. You'll be asked to specify the endpoints of the interval for plotting.
Print formula	Display the formula of the current function. This is particularly useful after building a complicated function.
Rescale	Change the plotting coordinates. This option works in the same way that the Rescale option in the General menu does.
Evaluate (Table)	Evaluate the current function for a given value or values of x. This option works just as the Evaluate option in the General menu does.
Clear	Clears the plotting window and "forgets" your current functions. Use this option when you want to begin a completely new example.
Exit	Return to the main menu bar.

We suggest that you try the saved examples on your disk as an introduction to Build Functions. See Figure 12. You may also wish to play the saved session COMPOSE to see examples of using it for composition of functions.

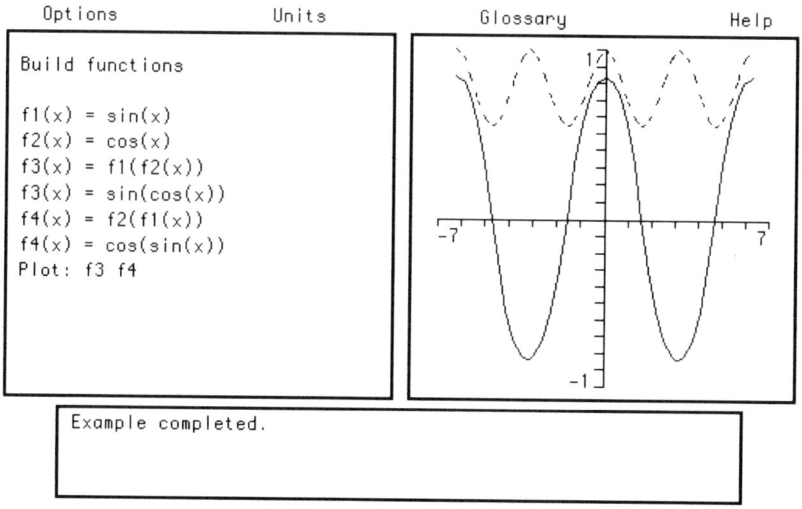

Figure 12

Using PRE-CALCULUS for Self-Study

This section includes some suggestions for use of PRE-CALCULUS in a laboratory setting, or for self-study by an individual student.

Checking answers

The simplest use of the package is to check answers on homework problems. For best results you should first work the problem on paper and then use the appropriate unit to check the answer(s). You may use any of the topics in this manner.

For example, your textbook exercises may ask you to evaluate a function at a particular point. You can check your answer by entering the function in the General menu, then choosing the option Evaluate.

When you are asked to sketch a graph of a function, do so with pencil and paper, then use the graphing capability of the General menu to compare your sketch with the actual graph of the function.

You'll be required to find roots of an equation in word problems as well as in straightforward exercise sets. Use the Roots topic to verify your answers.

The topic Trig Identities is ideal for checking your work in a unit on trigonometric identities. Once you have determined what you believe to be an identity, enter both sides of the equation into the Trig Identities dialog box and verify your answer.

Graphics

Most of the topics provide a graphic display. This helps to understand a problem, even if no graph is required for the solution.

For example, in finding roots of an equation, the software graphs the function so you can see by observation where the curve crosses the X-axis. For an even clearer understanding of the graphic interpretation of roots, try the option Solve graphically. That will give you a step by step visual display of root-finding.

When you're asked to find the inverse of a function, try the topic Inverse function. If the function has an inverse, you'll be able to compare graphically the function and its inverse. If the function does not have an inverse in a chosen interval, you can see by observation why not.

The plotting option in the topic Build functions provides a visual picture of functions built by composition. You can observe graphically why $f(g(x))$ is usually not the same as $g(f(x))$.

The plotting capabilities also give a clear picture of what it means when a function is undefined at a given point. Try plotting such functions as tan(x) from -*pi* to *pi* or (x-1)/(x^2-x-6) from -4 to 4.

Applications

Work more complicated applications by setting up the problem and using PRE-CALCULUS to find the solution.

Use pencil and paper to set up the functions or equations needed to solve a particular word problem, then use the appropriate topic in PRE-CALCULUS to do the computations and provide the final answer.

The Triangles menu is especially helpful in solving any problems which involve working with line of sight, angle of elevation, angle of depression, etc. Consider the following problem:

> *From a point on the ground 215 feet from the base of a tower, the angle of elevation of the top of the tower is 53 degrees. What is the height of the tower?*

First draw a picture which includes the given quantities and the quantity you want to determine. Remember that the angle the tower makes with the ground is 90 degrees, so you are given two angles and the included side. Use the option ASA to solve the triangle completely. The answer to the specific problem question will be the vertical side of the triangle.

Note that you can also use the Triangles menu to solve triangles which do not include right angles.

Use the General or Roots menus to solve problems involving compound interest, bacterial growth, radioactive decay, and applied problems in many other areas.

Use the Sine curves menu to examine problems in harmonic motion.

Deeper understanding

One of the main uses of the PRE-CALCULUS package is to give you a better understanding of the subject.

The form in which the software asks you to enter a function helps to reinforce the definition of a function. Note that the topics, such as General, that ask you for a function, request a single formula expressing *y* in terms of *x*.

Of course, you can't do this for a relation that isn't a function (for example, the equation of the circle $x^2 + y^2 = 4$).

This does not mean, however, that you can't graph a relation – it simply reinforces the fact that such relations are not functions. To graph a relation, such as a circle of radius 2, simply treat it as two functions:

f(x) = sqr(4 - *x*^2)

f(x) = -sqr(4 - *x*^2)

You can plot both of those functions and use the Rescale option to obtain the desired appearance.

Examine what happens when you rotate a function and its inverse to gain a better understanding of the meaning of an inverse.

Use the Build functions topic to clarify your understanding of composition of functions. Both the Print formula option and the Plot option can help you understand what happens to various functions under composition.

Free exploration

We hope that you will use your imagination for free exploration of topics in elementary functions and trigonometry. Here are a few suggestions.

Use the General menu to approximate the value of e. Enter the function (1 + 1/*x*)^*x* and evaluate for values of x. How large a value of x is necessary to obtain a good approximation?

Start with a function such as *f(x)* = 4^*x*. What if you change the function to 4^(2***x*)? What if the function were 4^(-2***x*)? Can you describe in words how the graph changes when you change the exponent?

Experiment with the topic Angles. For what angles are certain trigonometric functions not defined? Can you see from the graph why they are not defined?

For conic sections, there is a relation among A, B, and C that determines the type of conic. Try to discover what that relation is.

Try to find a trigonometric identity which relates cot(x) to csc(x).

Saved Sessions on your Disk

PRE-CALCULUS makes it very easy to record "sample sessions." If you're using PRE-CALCULUS to teach a class, you may want to create your own sessions to show in class.

Your disk also includes the following pre-recorded sessions. Each has been carefully designed to illustrate important ideas about the relevant topic.

GENERAL	Illustrates the General menu with examples of several functions
LOGS	Examples of logarithms to different bases using the General menu
ANGLES	Several examples using the Angles menu
TRIANGLE	Examples of solving triangles
LINES	Finding the equation of a line using two points and then one point and the slope

SINES	Illustrates the effects of changing parameters A, B, and C on the general function A*sin(B*x+C)
CONICS	Illustrates several conic sections
POLAR	Illustrates several examples of polar curves
ROOTS	Illustrates finding roots of equations
INVERSE	Illustrates examples of inverses of functions
IDENTITY	Verifies trigonometric identities
COMPOSE	Illustrates composition of functions

Index

Angles 2, 31-32
Apple Macintosh 5
Applications 61-62
Arithmetic operators 17
Backup copy 6
Build Functions 55-57
Built-in examples 13
Checking answers 59
Clear 26, 56
Color 9
Composition of functions 3, 55-57
Conic sections 2, 43-45
Constants 21
Degrees 11, 31
Dialog box 13
    editing 14-15
Editing keys 14
Ellipse 43
Entering functions 17-19
    examples 19
Equation of a line 33-34
Error message 19
Errors 28
    correcting 28
Evaluate 26, 56
Examples 19, 27-29, 40-41
    built-in 13
Free exploration 63-64
Functions 18
    building 55-57
    built-in 18
    composition 57
    entering 17-19
    inverse 51-52
    trigonometric 31
General menu 2, 29
Glossary 10
Graphics, in self-study 60-61
Hard disk 6
Help 10
Help menu 10
Hyperbola 43
IBM PC 5
Identities,trigonometric 3, 53-54
Inverse Functions 2, 51-52
Line style 9
Lines 33-34
Main menu 10
Menu 10
    glossary 10
    help 10
    plot 25
    pull-down 13
    session 10-11
    topics 10
    units 10
Menu choice 12
Message window 19
Networks 6
Numbers 21
    entering 21
Options menu 26, 33, 35, 39, 50, 55
Parabola 43
Parameters, with sine curves 39
Pi 18, 21, 50
Play a session 11
Plot menu 25
Plotting multiple functions 9
Polar Curves 2, 47-48
Print formula 56
PS/2 5
Quit 10, 12
Radians 11, 31
Record a session 1-2, 12
Replaying a session 1-2, 11-12

Requirements 5
Rescale 25, 40, 56
Roots 2, 49-50
Saved sessions 65
Self-study 59-64
Session menu 10-11
Sine Curves 2, 39-41
    examples 40
Slope 34
Solving triangles 35-37
Table 26, 56
Topics 23-57
Topics menu 10
Triangles 2, 35-37
Trigonometric functions 31-32
Trigonometric identities 3, 53-54
Units menu 10, 11
Windows 9
X-intercept 33
Y-intercept 33